Our World

Water Power

By Chris Oxlade

Aladdin/Watts

London • Sydney

© Aladdin Books Ltd 2006

Designed and produced by
Aladdin Books Ltd
2/3 Fitzroy Mews
London W1T 6DF

First published in 2006 by
Franklin Watts
338 Euston Road
London NW1 3BH

Franklin Watts Australia
Hachette Children's Books
Level 17/207 Kent Street
Sydney NSW 2000

A catalogue record for this
book is available from the
British Library.

ISBN 0 7496 6277 8

Printed in Malaysia

Editor:
Katie Harker

Designer:
Flick, Book Design and Graphics

Consultants:
Jackie Holderness – former Senior Lecturer
in Primary Education, Westminster
Institute, Oxford Brookes University.

Rob Bowden – education consultant,
author and photographer specialising in
social and environmental issues.

Illustrations:
David Burroughs, Mike Lacy,
Ian Thompson

Picture researcher:
Alexa Brown

Photocredits:
Abbreviations: l-left, r-right, b-bottom,
t-top, c-centre, m-middle
Front cover, 24br, 25bl – Digital Vision.
Back cover, 3tl, 3tml, 5ml, 5bl, 6tl, 6mr,
8tl, 9bl, 10tr, 10ml, 11tl, 16tl, 17tl, 18bl,
20bl, 21tl, 25tr, 26bl, 28tl, 29tr, 29b, 30tr,
30br – www.istockphoto.com. 1 – Select
Pictures. 2-3, 4ml, 12tl, 18tl – Comstock.
3bl, 5mbl, 6br, 13tl, 14tl, 27tr, 30mr – US
Department of Energy. 3bml, 20tr –
Courtesy of Marine Current Turbines Ltd.
4bl, 12bl, 16br – Itaipu. 5mr, 14br, 24tl,
31tl – Photodisc. 5br, 7bl, 22tr, 26tr, 31bl
– Corel. 7tr, 8br – Corbis. 15tr – Columbia
Power. 17br – Brad Templeton. 19tr –
Rainbow Power Company. 22bl –
Wavegen. 23tr – Ocean Power Delivery
Ltd. 27br – Portland General Electric. 28br
– Angelo Turconi.

CONTENTS

Notes to parents and teachers

This series has been developed for group use in the classroom as well as for children reading on their own. In particular, its differentiated text allows children of mixed abilities to enjoy reading about the same topic. The larger size text (A, below) offers apprentice readers a simplified text. This simplified text is used in the introduction to each chapter and in the picture captions. This font is part of the © Sassoon family of fonts recommended by the National Literacy Early Years Strategy document for maximum legibility. The smaller size text (B, below) offers a more challenging read for older or more able readers.

Water power today

Today, most of the water power that we use is in the form of electricity.

A

◀ Turbines have fins like the blades of a fan.

This turbine has fins (sometimes called vanes) that catch flowing water.

B

Questions, key words and glossary

Each spread ends with a question which parents and teachers can use to discuss and develop further ideas and concepts. Further questions are provided in a quiz on page 30. A reduced version of pages 30 and 31 is shown below. The illustrated 'Key words' section is provided as a revision tool, particularly for apprentice readers, in order to help with spelling, writing and guided reading as part of the literacy hour. The glossary is for more able

or older readers. In addition to the glossary's role as a reference aid, it is also designed to reinforce new vocabulary and provide a tool for further discussion and revision. When glossary terms first appear in the text, they are highlighted in bold.

 ## See how much you know!

What is the lake behind a dam called?

What makes tides rise and fall?

How can fish get past a dam on a river?

What does a tidal barrage do?

What do the generators do at a hydroelectric power station?

How does an overshot water-wheel work?

Name three good things about water power.

Name three problems that are caused by water power.

Key words

Dam

A

Electricity

Generator   **Reservoir**

River   **Tide**

Turbine   **Water**

Wave

Glossary

Barrage – A type of dam used at the coast to trap ocean water.

Cog – A wheel with teeth around the outside that is used to turn machinery.

Concrete – An extremely strong material made by mixing water, gravel and sand.

Decompose – To decay or rot.

Evaporate – To turn from liquid to gas.

Global warming – The gradual warming of the Earth's atmosphere. This is believed to be caused by the release of gases from rotting plants or activities such as burning fuels.

Hydroelectricity – Electricity generated by converting the energy of running water.

B

Oscillating – Going one way and then the other, again and again.

Particle – A very small piece of material.

Renewable energy – A source of energy that will never run out.

Trough – A long, narrow container with an open top, like a pipe cut in half.

What is water power?

Water power is the energy that we get from flowing water. We use this energy to make electricity for our homes, schools, shops and factories. Electricity made from flowing water is called hydroelectricity.

► **A dam is used to make hydroelectricity.**

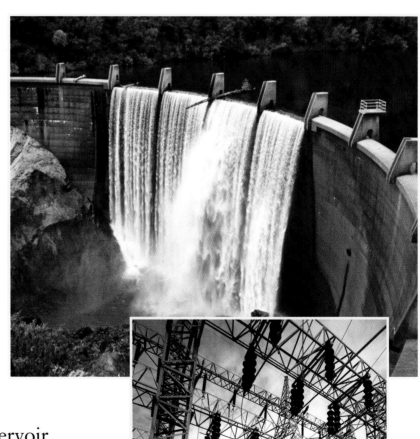

A **hydroelectric** power station normally has a large wall called a dam. The dam is built across a river. It traps flowing water, which builds up to make a lake called a reservoir. The water flows from the reservoir to the power station's power house, where its energy is turned into electricity.

▶ Flowing water contains lots of energy.

Anything that is moving has energy in it. Hydroelectric power stations capture the energy in flowing water and change it into electricity. Electricity is a type of energy too.

Water power makes about a sixth of the world's electricity.

We use this electricity for heating and lighting. Water power is often used in countries that have high mountains or steep valleys. These features cause fast flowing river water that has lots of energy in it. Some countries are very dependent on hydroelectricity. They get nearly all of their electricity from flowing water.

 What force (push or pull) makes water flow downhill?

The water cycle

Most water power comes from water flowing down rivers. Rivers carry water from the land to the sea. This forms part of a process called the water cycle. Without it we would have no water power.

▶ **Clouds are made from tiny water drops.**

Clouds form when **particles** of water in the air clump together to form millions of tiny water drops or small crystals of ice. The water particles come from the surface of the sea and from wet places and plants. If the water drops get big and heavy enough they fall from the air, making rain, snow or hail.

▼ Rivers carry water back to the sea.

Rain-water runs off the land and into streams. The streams join together to make small rivers. The small rivers join together to make larger rivers. Eventually the rivers reach the sea. The water that started in the sea eventually returns to the sea.

Water moves in a constant cycle.

Heat energy from the Sun makes water **evaporate**. The water drops move into the air and cool to make clouds. When clouds release their rain over land, the water falls to the ground and is carried by streams and rivers back to the sea.

Water falls as rain.

Water vapour rises.

Water vapour evaporates.

Rivers flow downhill to the sea.

 What types of energy help the water cycle to work?

Water power in the past

People have made use of water power for the last 4,000 years. We know this because we have found old water-mills. The first water-mills were used to make flour or to weave cloth.

▲ **This water-wheel turns.**

A water-wheel has buckets or paddles around the outside. Water flows from a river or millpond along a **trough** to the wheel. As the water flows, it fills the buckets or pushes against the paddles. The force of the water makes the wheel turn. The water then joins the river again on the other side.

◀ **These machines are worked by a water-wheel.**

The machinery in this water-mill is used to weave cloth. The water-wheel outside the building turns a rod inside the building (A). The rod is connected to the weaving loom by large **cogs** (B). These looms are used to weave materials such as cotton and silk.

Water can flow under the wheel or over the wheel.

There are two main types of water-wheels – overshot wheels and undershot wheels. In an overshot wheel, water flows slowly to the top of the wheel. It fills buckets or pushes paddles on one side which makes the wheel turn. In an undershot water-wheel, water flows quickly under the wheel. It catches the buckets or paddles to turn the wheel.

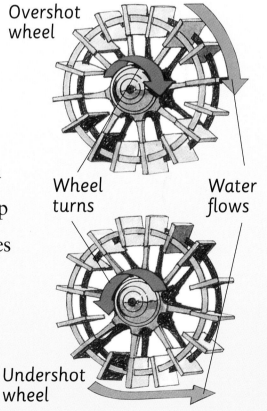

Overshot wheel

Wheel turns

Water flows

Undershot wheel

 What other jobs have water-wheels been used for?

Water power today

In some places, people still use water-wheels to turn machinery. Today, however, most of the water power that we use is in the form of electricity. Flowing water turns a turbine which then turns a generator to make electricity.

◀ **Turbines have fins like the blades of a fan.**

This turbine has fins (sometimes called vanes) that catch flowing water. The flowing water pushes on the fins, just as water rushing from a hose can push things along the ground. The push makes the turbine spin. The faster the water flows, the faster the turbine spins.

◀ A generator makes electricity when it turns.

A generator contains hundreds of coils of wire. Some of these coils are attached to a shaft in the middle. Some are round the outside of the shaft. When a turbine spins, the generator turns this movement into electricity and it flows through the coils.

Water flow

Turbines are used to capture the power of flowing water.

In this turbine, water flows through the top of the turbine and downwards, like water going down a giant drain. As it does, it catches the turbine's fins and makes them spin very fast.

This turbine uses water that enters from the side, and catches the turbine's fins as it flows down. At a large power station each turbine is the size of a house!

Water flow

What do we use electricity for?

Hydroelectric dams

River water is used in a hydroelectric power station. A dam is used to trap the water before it flows into the power-house. Here it works the turbines, before returning to the river below the dam.

► Some dams are over 150 metres tall.

Dams can also be several kilometres long. A dam has strong **concrete** walls to hold the deep water behind. If the reservoir becomes full and is in danger of flooding, the extra water flows down a spillway. This also regulates the amount of water that the power station uses.

◀ This power-house is being built in a cave.

The turbines and generators of a hydroelectric power station are in a giant space called a power-house. This is sometimes a building at the bottom of the dam, sometimes inside the base of the dam and sometimes in a giant cave in the rock beside the dam.

This diagram shows the parts of a power station.

Water from the reservoir flows down huge pipes to the power-house. The amount of water flowing down is controlled by valves, like giant taps.

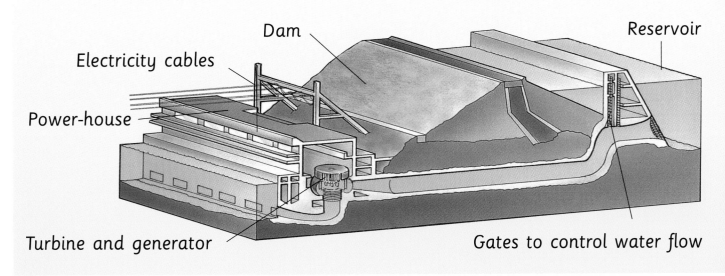

Electricity cables

Dam

Reservoir

Power-house

Turbine and generator

Gates to control water flow

 What would happen if water flowed over the top of a dam?

Building giant dam

Dams must be extremely strong to hold back reservoir water at a hydroelectric power station. Engineers must choose carefully where to build a dam. They need river water and a valley for the reservoir.

► **Concrete dams are very strong.**

The Itaipu Dam in South America has extremely thick walls of reinforced concrete. These walls hold back the force of the Paean river. The reservoir behind the dam is 1,350 square kilometres. Over 30,000 people took seven years to build the dam.

◀ In steep valleys, dams are made into an arch.

The type of dam that engineers build depends on the width and shape of the valley, the height the dam will be and the type of rocks. The Hoover Dam in the US is an arch concrete dam. The wall is thin, but it is super strong because it is curved, like an arched bridge on its side.

Some dams take over ten years to build.

Building a dam is a massive project. First the area must be cleared and the building materials transported. Then the river must be diverted while the dam is being built.

▶ The Three Gorges Dam is being built on China's largest river, the Yangtze (Chang Jiang). The project started in 1993. On completion in 2009, the dam will provide electricity to Shanghai and six nearby provinces.

 What different materials are used to make a dam?

Micro water power

Large hydroelectric power stations produce enough electricity for whole cities. But some smaller stations are also used. Very small power stations are called micro hydroelectric schemes. They make electricity for a village, a small factory, a farm or even a house.

◀ **Some small water power stations do not need a dam.**

Instead of using a dam, some small hydroelectric schemes use river water that flows into a pipe or a trough. The water in the pipe or trough flows to a small turbine and generator. These power stations do not alter the landscape as much as power stations that use dams.

▶ This is the turbine in a small water power station.

Small and micro hydroelectric schemes have a turbine called a pelton wheel (right). The wheel has a rim with cups or ridges. As fast-flowing water hits these, the wheel spins and turns a generator to make electricity.

Water jet turns pelton wheel.

This power station makes energy for just one house.

The generator makes electricity that goes to the mains electricity circuits in the house. When no power is needed in the house, the electricity from the generator is stored in the batteries. The batteries also supply electricity if the turbine or generator break down.

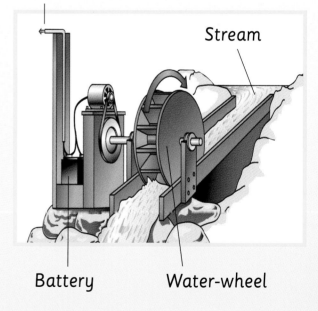

Power to house

Stream

Battery   Water-wheel

  Why would it be a bad idea to rely on hydroelectricity alone?

Power from tides

Water also flows in the oceans. Ocean tides rise and fall and waves break on the coast, too. Special machines, such as tidal turbines (right), are used to catch the energy from ocean water.

◀ **The tide goes in and out because the Moon pulls on the ocean water.**

The Moon's gravity pulls on the ocean water. As the Earth spins round, this pull causes the water to slowly slosh about. This makes the level of the water rise and fall on the coasts around the world. In most places the tide goes in and out twice a day. Tide energy can be used again and again. We call this **renewable energy**.

▲ **Tidal power stations can be built at the coast.**

Bays or river mouths on the coast are a good place for tidal power stations. A dam called a **barrage** is used to trap the water, which then flows through the barrage, turning a turbine and a generator inside. The world's biggest tidal power station is being built in South Korea and due to be completed in 2009. It will provide electricity from the constant flow of water in and out of a seaside bay.

A tidal power station uses flowing water to make electricity.

As the tide rises, water moves into the power station. The force of the water turns turbines, which are used to generate electricity. Water can also be stored for use at low tide.

Water is stored for use at low tide.

Water enters as the tide rises and leaves as the tide falls.

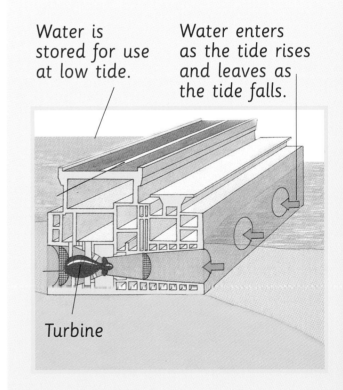

Turbine

 In what ways can tidal power be used again and again?

Power from waves

Waves are made when the wind blows across the sea. The stronger the wind, the taller the waves get. A wave power machine catches the energy in this movement. This is called wave power.

◀ **This wave power station is on the coast.**

This type of wave power station is built on rocks on the seashore. As waves crash into the shore, they force air in and out of a column inside the power station. This spins a turbine and works a generator to make electricity. The power station's design makes it easy to build and install, and it doesn't impact too much on the coastal landscape. Other types of wave power station can be used out at sea, away from view.

▶ This wave power machine makes electricity as it bends in the waves.

This machine works at sea and is called a Pelamis. As waves go by, it bends up and down. This motion works pumps, which turn a turbine and a generator.

In a wave power station, the moving water pushes air that makes a turbine spin.

This power station is an **oscillating** water column (OWC). It has a large tube with its bottom end submerged in the water. Waves make the column of air in the tube move up and down (1) as waves move into (2) and out of (3) the tube. Air rushes in and out of the top of the tube (4), which spins a turbine and works a generator.

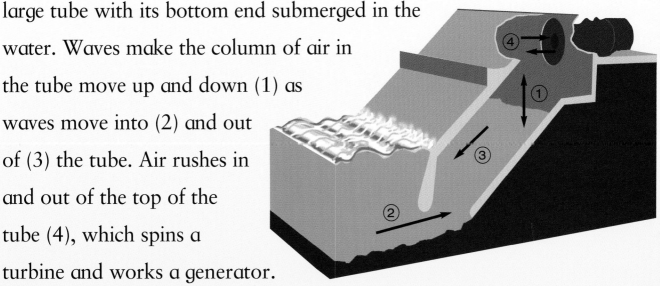

 Is wave power a good way to produce electricity all the time?

Advantages of water power

Why is hydroelectricity better than other forms of power? Water can be used again and again to make power and it doesn't burn fuels that release polluting gases. The dams of power stations also help to reduce flooding and provide reservoirs for us to use.

▶ **Water power does not burn fuels like this coal power station.**

Most power stations burn coal to make electricity. Burning coal makes gases, such as carbon dioxide and sulphur dioxide, which pollute the air. Water power has another advantage over coal power stations. Although coal will run out, water is renewable – it is constantly moving in a cycle.

▶ Water from reservoirs can be used for drinking and washing.

The reservoirs of water power stations store water when there is lots of rain. This water is piped to homes, offices, factories and farms where it is used for drinking, washing, watering crops and for making things.

Big dams help to stop floods.

Floods happen after heavy rain when rivers burst their banks. Reservoirs can catch the sudden rush of water coming down a river and then release it slowly. This stops flooding downstream of the dam. But dams have also collapsed in the past, creating floods instead of stopping them.

What else can we use reservoirs for?

Water power problems

Water power seems to be the perfect 'clean' source of energy. But dams cause many problems, too. They stop fish and other animals moving up and down rivers. They slow down boats. A dam reservoir can also flood land and submerge houses.

▼ These trees have been flooded by a reservoir.

New dams can spoil the landscape. Huge areas of land are flooded to make a reservoir. Trees and other plants are killed and people and animals are forced to move to other areas. When plants are submerged by flood water they rot away. As they **decompose**, the plants release methane, a gas that contributes towards **global warming**.

▶ A fish ladder helps fish to swim past a dam.

Some dams have a series of pools up the side called a fish ladder. Fish can pass the dam by swimming or leaping up the low steps. The flow of water must be fast enough to attract the fish, but not too fast to wash them back downstream again. Sadly, many fish don't find the swim ladder and are injured or killed in the turbines.

This dam will be knocked down in 2007.

Many countries have stopped building dams because of the problems they cause. The Marmot dam in the US (right) is being taken down because it harms salmon, a threatened species. Other old dams are being destroyed because they leak and cause flooding.

 What other reasons might be given for removing a dam?

Water power in the future

As the world's population grows, we will need more electricity in the future. There are not enough places to build dams to make all the electricity we will need, but water power is always likely to be used to make some of it.

▶ **This river in Africa will power electricity.**

There are plans to improve hydroelectric facilities at Inga (rapids found on the Congo river) and to build a new power station. These projects will provide twice the electricity of the Three Gorges Dam – enough to meet the whole of Africa's annual electricity demand.

► Water pipes like these can also generate energy.

In some towns in the US, hydroelectric generators are used in the water mains to produce electricity that can be used locally.

Warm water

Cold water

Cold water turns gas back into a liquid again.

Warm water changes liquids into gases that can be used to generate electricity.

Cold water

Changes in water temperature can be used to make energy.

The ocean's surface traps heat from the Sun but deeper parts of the sea are cooler. Scientists use these temperature changes to turn liquids into gases that can be used to drive turbines and generate electricity. This is called ocean thermal energy conversion (OTEC).

Thermal pools, like those found in Iceland (left), can also be used to generate heat and electricity. In Iceland, houses are heated by thermal water pumped from the ground.

Where does the heat in ocean water come from?

See how much you know!

What is the lake behind a dam called?

What makes tides rise and fall?

How can fish get past a dam on a river?

What does a tidal barrage do?

What do the generators do at a hydroelectric power station?

How does an overshot water-wheel work?

Name three good things about water power.

Name three problems that are caused by water power.

Key words

Dam

Electricity   Energy

Generator   Reservoir

River   Tide

Turbine   Water

Wave

Glossary

Barrage – A type of dam used at the coast to trap ocean water.

Cog – A wheel with teeth around the outside that is used to turn machinery.

Concrete – An extremely strong material made by mixing water, gravel and sand.

Decompose – To decay or rot.

Evaporate – To turn from liquid to gas.

Global warming – The gradual warming of the Earth's atmosphere. This is believed to be caused by the release of gases from rotting plants or activities such as burning fuels.

Hydroelectricity – Electricity generated by converting the energy of running water.

Oscillating – Going one way and then the other, again and again.

Particle – A very small piece of material.

Renewable energy – A source of energy that will never run out.

Trough – A long, narrow container with an open top, like a pipe cut in half.

Index